Jobs People Do

Meteorologists

by Emily Raij

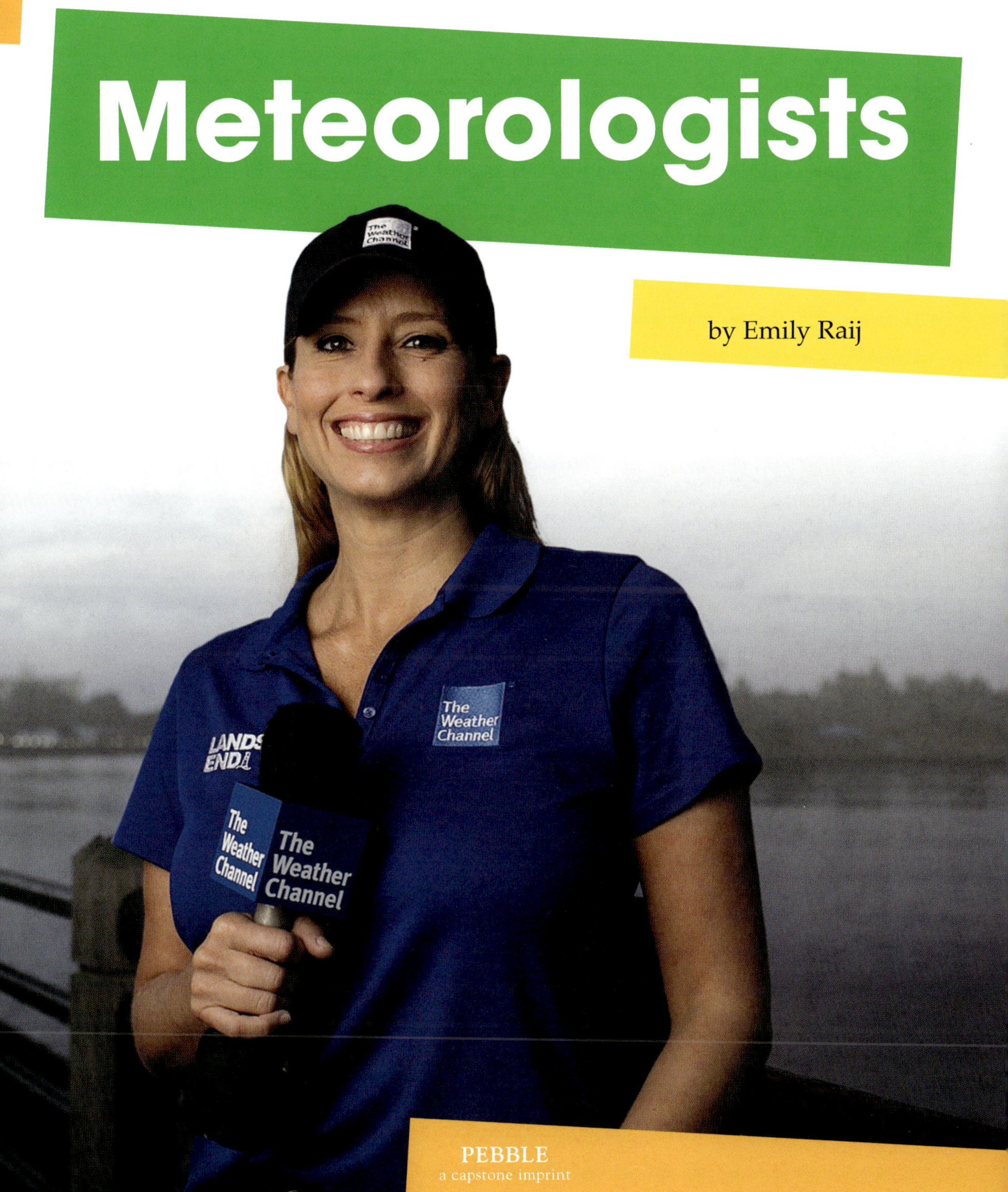

PEBBLE
a capstone imprint

Pebble Explore is published by Pebble, an imprint of Capstone.
1710 Roe Crest Drive, North Mankato, Minnesota 56003
www.capstonepub.com

Library of Congress Cataloging-in-Publication Data is available on the Library of Congress website.
ISBN: 978-1-9771-1377-1 (library binding)
ISBN: 978-1-9771-1812-7 (paperback)
ISBN: 978-1-9771-1385-6 (eBook PDF)

Summary: Meteorologists study the weather and tell people what it will be like. Learn all the roles meteorologists have, the tools they use, and how people get this exciting, fast-paced job.

Image Credits
Alamy: Dennis MacDonald, 18-19; iStockphoto: Images_By_Kenny, 8, simonkr, 20; Library of Congress Prints and Photographs, 28; LLC via AP Images/Weather Group Television/Justin Kase Conder, 1; Newscom: Blend Images/Noel Hendrickson, 23, MEGA/Joe Russo, 5, Xinhua News Agency, 11, ZUMA Press/Michael A. Jones, 9, ZUMA Press/NOAA, 16; Science Photo Library, 15; Science Source: Howard Bluestein, 12-13, Jim Reed Photography, 7, 17; Shutterstock: Andrey Burmakin, Cover, imagehd, 21 (radar), Mega Pixel, 21 (air thermometer), Mette Fairgrieve, 21 (weather balloon), NASA Images, 21 (satellite pictures), NATALINOSOVA, 14, 21 (wind speed meter), Nejron Photo, 25; U.S. Navy photo by John F. Williams, 27, Mass Communication Specialist 2nd Class Ryre Arciaga, 21 (computers)

Editorial Credits
Editor: Carrie Sheely; Designer: Kyle Grenz; Media Researcher: Jo Miller; Production Specialist: Kathy McColley

Printed and bound in the USA.
PA99

Table of Contents

What Is a Meteorologist?........................4

Types of Meteorologists...........................6

Where Meteorologists Work..................8

What Meteorologists Do....................... 10

How to Become a Meteorologist....................................... 24

Famous Meteorologists 28

Fast Facts.. 29

Glossary 30

Read More....................................... 31

Internet Sites 32

Index... 32

Words in **bold** are in the glossary.

What Is a Meteorologist?

Snow fell all night. It's still coming down fast. Snow covers rooftops. It sits on cars. The roads aren't clear. It's hard to see. Will there be school today? Click! The meteorologist on TV will know.

Meteorologists are **weather** scientists. They study the weather. Will the weather be cold or hot? Will it be snowy or rainy? Meteorologists tell others what the weather will be like in their **forecasts**. With them, you can plan your day!

A TV meteorologist gives a storm forecast.

Types of Meteorologists

Meteorologists do their work in different ways. Some tell what the weather will be like in a short time. They might say what the weather will be like in a day or in a week.

Others study weather **patterns** over many years. They look for changing patterns. Changing patterns can bring weather changes. The weather might be hotter.

Some do research. They study one part of weather science. They might study storms.

Meteorologists study a storm in Nebraska.

Where Meteorologists Work

Many weather scientists work for groups that the U.S. **government** runs. One is the National Weather Service (NWS). These workers give weather forecasts. They do research too. They look at weather in many places.

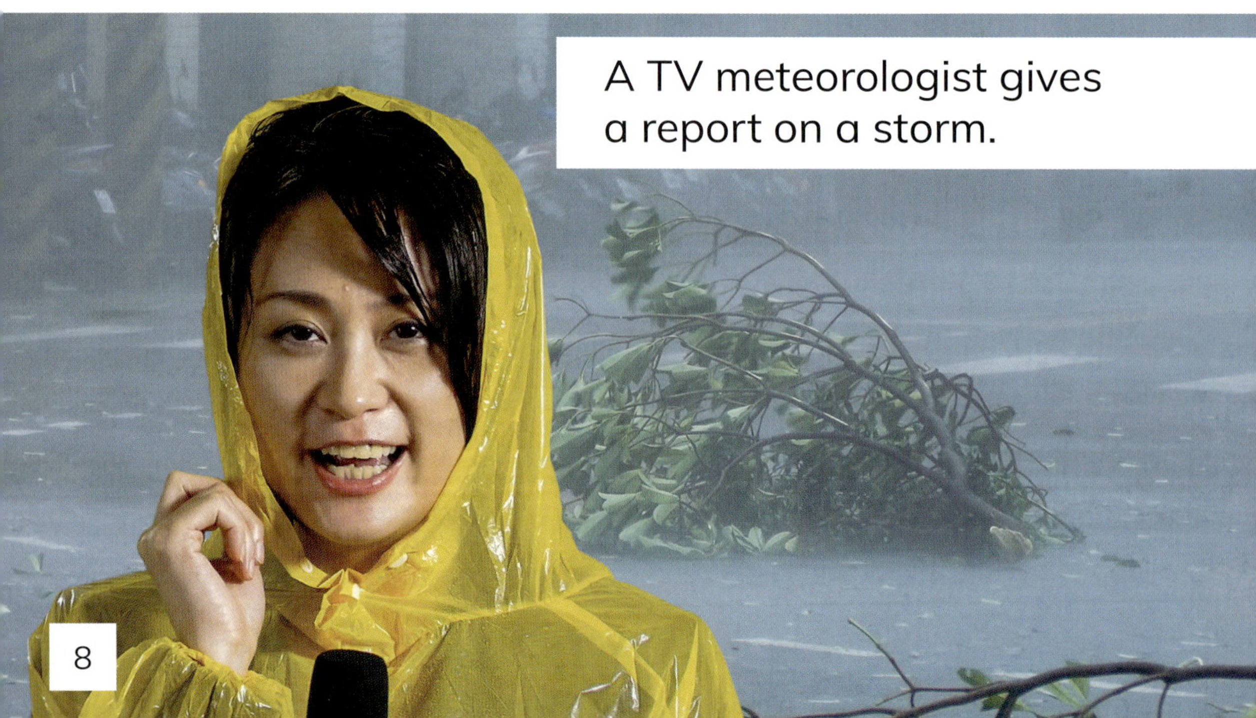

A TV meteorologist gives a report on a storm.

A meteorologist for the National Weather Service tracks weather on a computer.

Some meteorologists work at TV and radio stations. They tell people watching or listening the forecast.

Others work at **colleges**. They teach weather science classes.

What Meteorologists Do

Meteorologists study weather in different places. They get facts for their jobs. Sometimes they travel. They might go to cold places at the North and South Poles. The poles are at the top and bottom of Earth.

Weather scientists look at ice at the poles. They see how much is there. They see how much is melting. These facts help with forecasts. The weather at the poles can cause weather changes in many other places.

A team sets up tools to study weather at the South Pole.

Some weather scientists follow dangerous weather. They go where storms are. They set up tools. Some show wind speed. Others show which way wind is blowing. Meteorologists look at what the tools show them.

A meteorologist uses a tool to study a tornado.

They learn how storms start. They see how storms grow. They may see which storms can make tornadoes. These strong, swirling winds touch the ground. They can wreck anything in their paths.

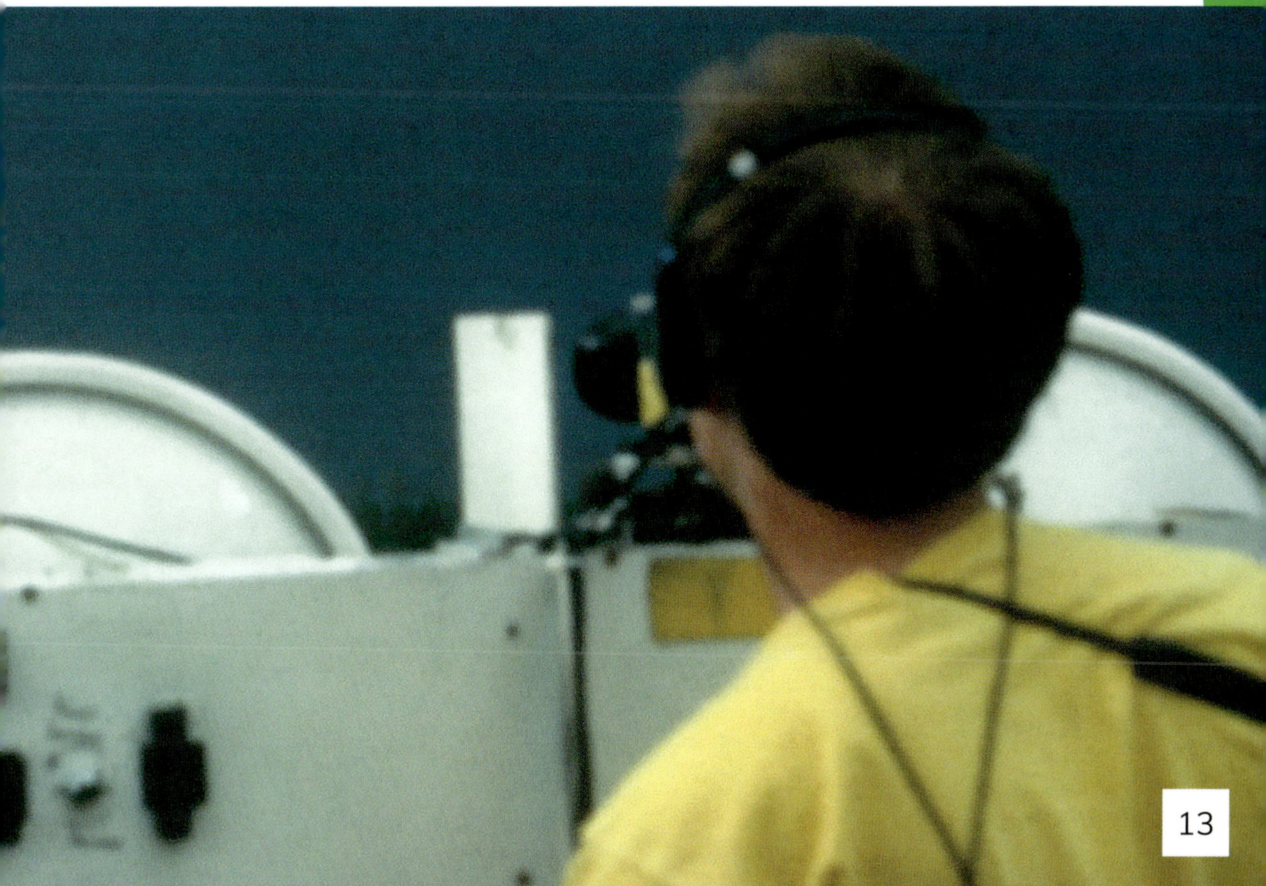

Meteorologists use many tools. Thermometers check air temperature. A wind speed meter checks the wind speed. Another tool shows how much water is in the air. If the air has a lot of water, rain might be coming.

Scientists release weather balloons into the air. The balloons check for changes as they go up.

wind speed meter

A meteorologist releases a weather balloon.

Weather scientists also use tools that show pictures. **Satellites** travel around Earth. They show pictures of clouds. They show where some storms are. They can see how much snow is on the ground.

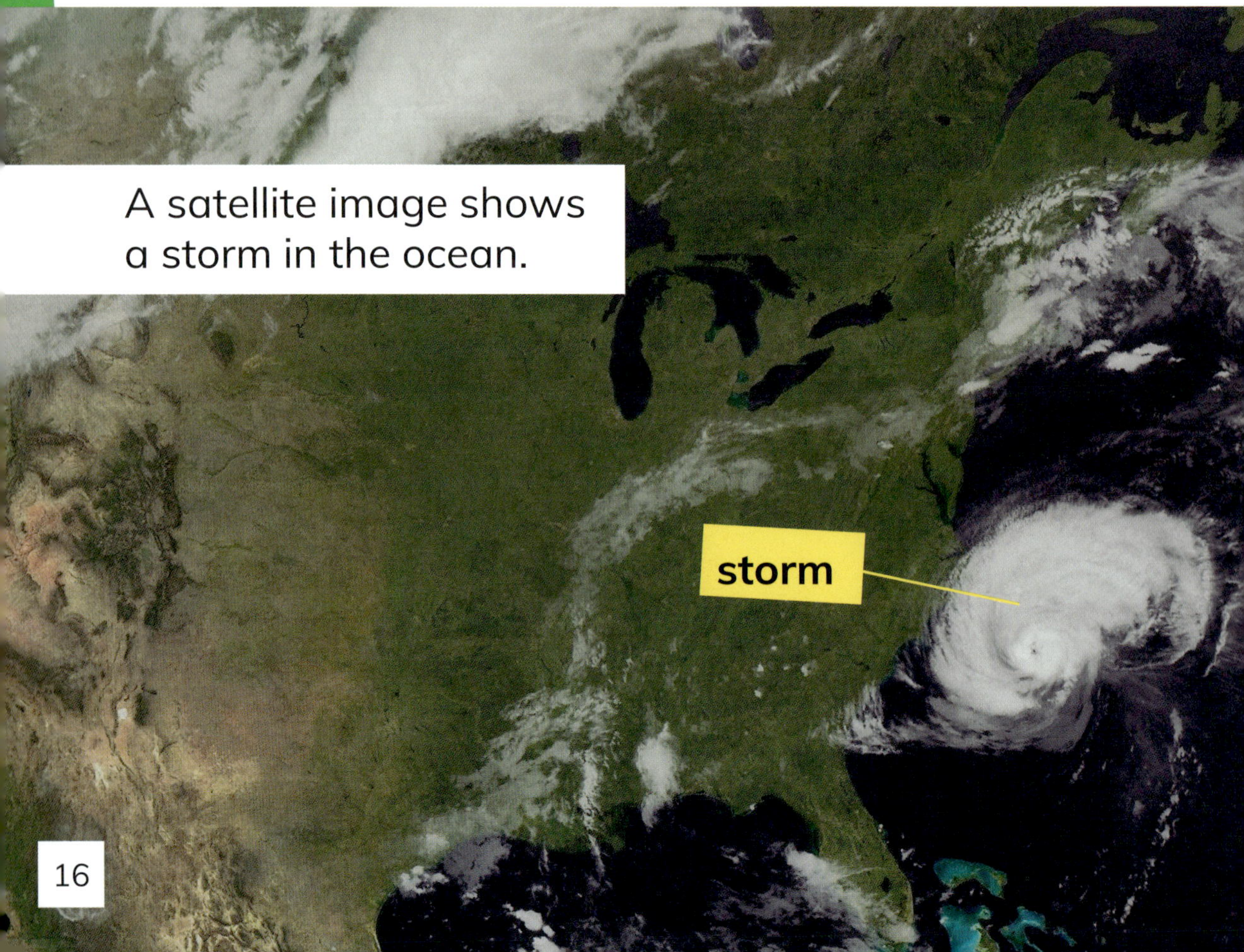

A satellite image shows a storm in the ocean.

Meteorologists look at radar pictures.

Radar tracks water that falls from the clouds. This water is called **precipitation**. It can be rain, snow, or other types. Radar shows where the water is going. Meteorologists keep track of storms with radar.

Weather scientists use computers too. They look at information from tools. They put this information into special computers. The computers show pictures. These are called **models**.

Scientists look at the pictures. They think about what they know about the weather. Then they make their forecasts. They might say it is going to rain or snow. They might say it is going to be hot or cold.

TV meteorologists show people watching TV the weather. They point to big maps. The maps show where it's raining and snowing. Some show air temperatures. They also show cold and warm air. These maps help people know what the weather will be where they live.

A TV meteorologist uses a map.

Tools Meteorologists Use

computers

radar

satellite pictures

wind speed meter

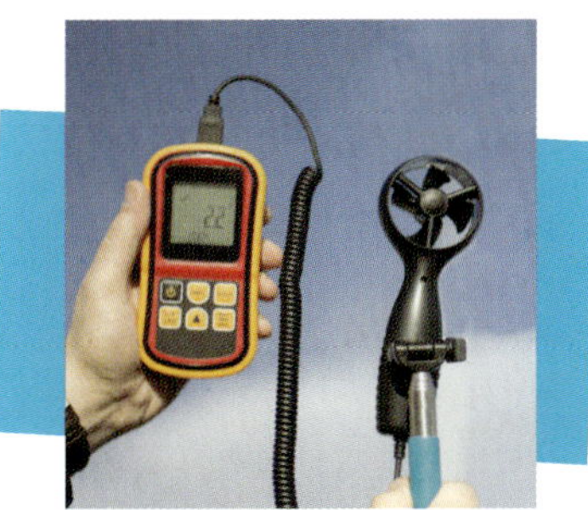

air thermometer

weather balloon

Meteorologists help people. Their forecasts might tell people it is going to rain all day. This can help people plan their day.

Meteorologists might see a storm coming. Then they can tell people ways to stay safe. They might tell people to get extra food and water. They can say when to find a safe place to stay. They tell people not to go outside or drive.

How to Become a Meteorologist

People who want to be meteorologists go to a college. They study for four years to get a **degree**. Students take many math and science classes. They take computer classes. They also learn how to use weather tools.

Some students learn the ways people make air dirty. They learn how it can cause weather changes over time.

Some weather scientists want to teach or do research. They often need to go to school longer. They might study for two to seven more years. They get another degree. This helps them learn one area of weather really well.

Students can also train with working meteorologists. They learn what the job is like.

Famous Meteorologists

There are many famous weather scientists. Cleveland Abbe was from the United States. He gave the first U.S. weather report about 150 years ago in 1871.

In 1970, June Bacon-Bercey became the first female TV meteorologist in the United States.

Cleveland Abbe

Fast Facts

- **What Meteorologists Do:**
  Meteorologists are weather scientists. They study the weather. They give forecasts.

- **Where Meteorologists Work:**
  U.S. government groups, TV and radio stations, colleges

- **Key Tools:**
  thermometers, weather balloons, weather radar pictures, satellite pictures, computers, wind speed meters, computer models, maps

- **Education Needed:**
  four years of college; sometimes up to seven more years for an advanced degree

- **Famous Meteorologists:**
  Cleveland Abbe, June Bacon-Bercey

Glossary

college (KOL-ij)—a school that students go to after high school

degree (di-GREE)—proof of graduating from college

forecast (FOR-kast)—a report of future weather conditions

government (GUHV-urn-muhnt)—the group that makes laws for a country

model (MOD-uhl)—a computer picture that shows future weather patterns

pattern (PAT-urn)—things repeated in the same way many times

precipitation (pri-sip-i-TAY-shuhn)—water that falls from clouds in the form of rain, hail, or snow

satellite (SAT-uh-lite)—an object in space that moves around a larger object to collect information

weather (WE-thur)—the conditions outdoors at a certain time and place

Read More

Braun, Eric. *Curious Pearl Dives Into Weather: 4D An Augmented Reading Science Experience*. North Mankato, MN: Picture Window, 2019.

de Seve, Karen. *National Geographic Little Kids First Big Book of Weather*. Washington, D.C.: National Geographic, 2017.

Roker, Al. *Al Roker's Extreme Weather: Tornadoes, Typhoons, and Other Weather Phenomena*. New York: HarperCollins, 2017.

Internet Sites

Earth Networks: Kid Meteorologist: How to Turn a Passion into a Career
https://www.earthnetworks.com/blog/kid-meteorologist/

PBS LearningMedia: Kid Meteorologist
https://www.pbslearningmedia.org/resource/ess05.sci.ess.watcyc.kidmeteor/kid-meteorologist/#.XV169pNKh24

PLAN!T NOW: Young Meteorologist Program
http://youngmeteorologist.org/about-ymp/kids/

Index

Abbe, Cleveland, 28

classes, 9, 24
colleges, 9, 24

degrees, 24, 26

forecasts, 4, 8, 9, 10, 19, 22

government, 8

National Weather Service (NWS), 8

patterns, 6
poles, 10

radio stations, 9
rain, 4, 17, 19, 20, 22
research, 7, 8, 26

snow, 4, 16, 17, 19, 20
storms, 7, 12–13, 16, 17, 22

tools, 12, 14, 16, 18, 24
  maps, 20
  models, 18
  radar, 17
  satellites, 16
  thermometers, 14
  weather balloons, 14
  wind speed meters, 14
tornadoes, 13
TV stations, 9